The Price of a Pioneer Journey

Adding and Subtracting Two-digit Dollar Amounts

Barbara M. Linde

PowerMath™

The Rosen Publishing Group's
PowerKids Press™
New York

Special thanks to Kathi McMahon and the staff of the
End of the Oregon Trail Interpretive Center, Oregon City, Oregon, for their assistance.

Published in 2006 by The Rosen Publishing Group, Inc.
29 East 21st Street, New York, NY 10010

Book Design: Michael J. Flynn

Photo Credits: Cover © Foodpix; p. 5 © Charlie Borland/Index Stock; p. 7 (store) © Mike McClure/Index Stock;
p. 7 (cans) © Scott T. Smith/Corbis; pp. 9 (both), 11 © Corbis; p. 13 © Bettmann/Corbis.

ISBN: 1-4042-3337-7

Library of Congress Cataloging-in-Publication Data

Linde, Barbara M.
The price of a pioneer journey / Barbara M. Linde.
p. cm. – (The Rosen Publishing Group's reading room collection. Math) Includes index.
ISBN 1-4042-3337-7 (lib. bdg.)
1. Arithmetic–Juvenile literature. 2. Pioneers–United States–Juvenile literature. I. Title. II. Series.
QA115.L6826 2006
513–dc22
2005012212

Manufactured in the United States of America

Contents

Going West!

In the 1840s, many people in the United States decided to move west to start a better life. These people were called **pioneers**. The pioneers formed groups to travel west together in wagon trains. They traveled about 2,000 miles in covered wagons pulled by horses, **oxen**, or mules. One of the main roads to the West was known as the **Oregon Trail**. It started from 2 cities in Missouri: St. Joseph and Independence. The trail then crossed through what are now the states of Kansas, Nebraska, Wyoming, Idaho, and Oregon.

Some pioneers bought a covered wagon to make the trip west. Others bought a farm wagon and turned it into a covered wagon.

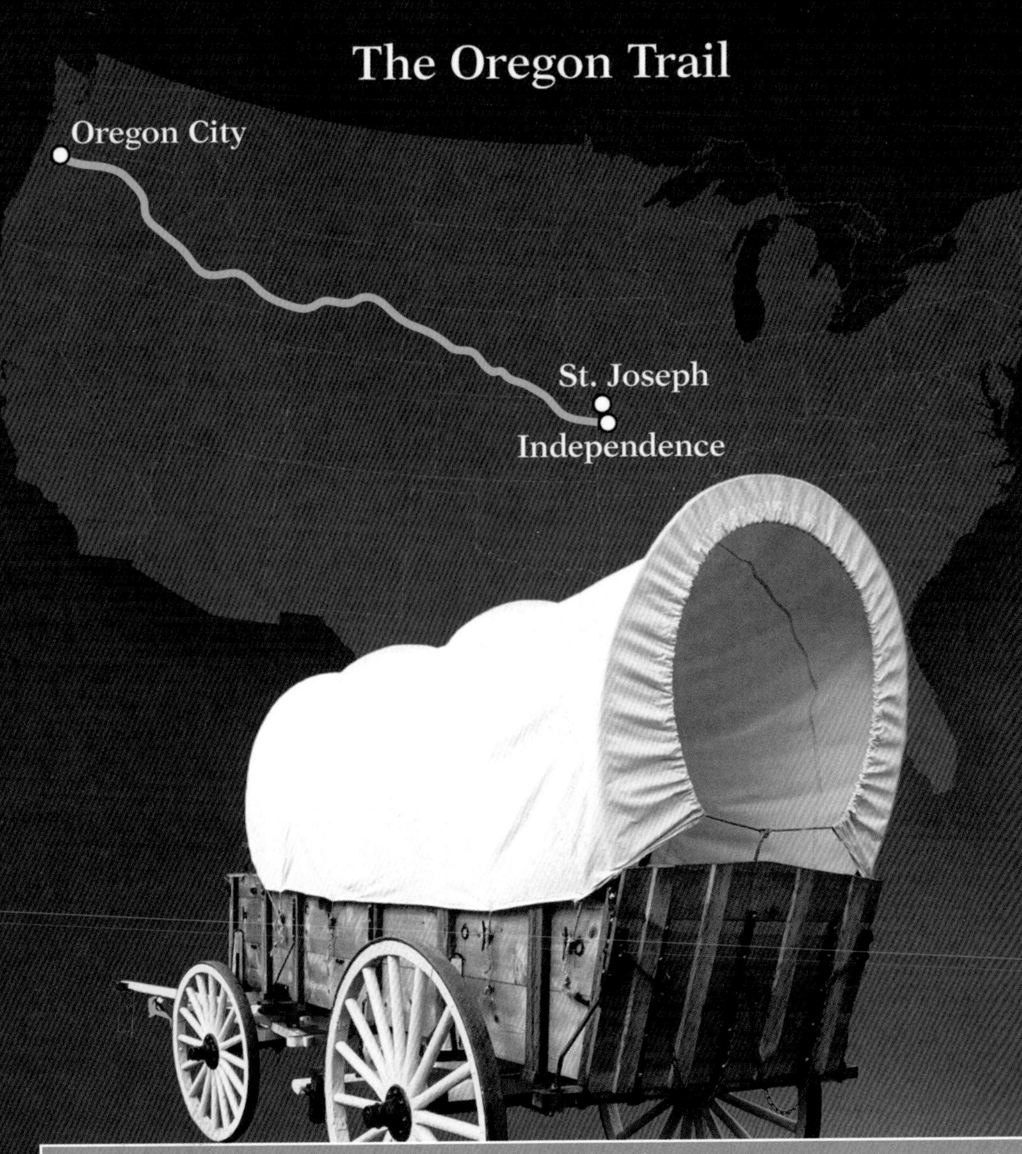

$30.00 + 18.00 $48.00	A farm wagon cost $30.00. The cloth and other parts needed to turn it into a covered wagon cost about $18.00. Add $30.00 to $18.00 to find out how much it cost altogether. Don't forget to use dollar signs and decimal points when you're adding or subtracting money!

Supplies for the Trip

Before the pioneers started their long trip
est, they had to buy animals and **supplies**. They
sed **addition** and **subtraction** to figure out how
uch money they would have to spend. They had
buy animals like oxen, cows, and horses. They
ad to buy things to eat like dried fruit, rice, flour,
nd beans. They had to buy pots to cook their
od. The pioneers sometimes bought all their
ipplies at one store, called a general store.

$$\begin{array}{r} \$97.00 \\ -\ 86.00 \\ \hline \$11.00 \end{array}$$

One pioneer had \$97.00. She spent \$86.00 for food. How much money did she have left? Subtract \$86.00 from \$97.00 to find out.

inside of a general store

On the Trail

The wagon trains had to cross rivers on the trip. Sometimes there would be a bridge the pioneers could use to cross the river. The people who built the bridge made the pioneers pay for each wagon that crossed it. Other people ran **ferries**, which were flat boats that could take a few wagons across the river. The pioneers had to pay to use the ferries, too.

$$\begin{array}{r} \$10.00 \\ +\ \ 60.00 \\ \hline \$70.00 \end{array}$$

A wagon train with 20 wagons paid $10.00 to cross a bridge. Then it cost them $60.00 to cross a river on a ferry. How much did they spend in all? Add the 2 amounts to find out.

bridge
ferry

Fort Laramie

After traveling for about 6 weeks, the pioneers got to Fort Laramie, in the area that is now Wyoming. They were able to rest and clean up at the fort. There was also a store at the fort where the pioneers could buy more supplies. After they left the fort, the pioneers crossed the Rocky Mountains. There were no more stores for hundreds of miles.

$$\begin{array}{r} \$35.00 \\ -\ \ 13.00 \\ \hline \$22.00 \end{array}$$

One family spent a total of $35.00 at Fort Laramie. They bought flour for $13.00. How much did they spend on other supplies? To find out, subtract the amount they spent on flour from the total amount.

Fort Laramie

The End of the Trail

After the pioneers got over the Rocky Mountains, they soon arrived at Oregon City, which was the end of the trail. Oregon City was near where the city of Portland, Oregon, is today. The pioneers needed new clothes and food. Some of the settlers who had reached Oregon earlier had opened stores. The new settlers were able to buy supplies and start their new lives in the West.

$$\begin{array}{r} \$67.00 \\ -\ 33.00 \\ \hline \$34.00 \end{array}$$

One pioneer spent $67.00 for supplies to build a cabin. He bought lumber for $33.00. How much did the pioneer spend for other supplies? To find the answer, subtract $33.00 from $67.00.

Oregon City

Addition and Subtraction

The pioneers bought many supplies. You are not a pioneer, but you buy things, too!

Let's say you got $40.00 for your birthday. You spend $10.00 on a book and $20.00 on a computer game. How much money did you spend? Use addition to find out.

How much money do you have left over? Use subtraction to find out.

Now you know how to use addition and subtraction to figure out what things cost and how much money you have, just like the pioneers did!

Glossary

addition (uh-DIH-shun) The adding of numbers to get a total.

ferry (FAIR-ee) A flat boat used to carry people, animals, and wagons across rivers.

Oregon Trail (OR-ih-guhn TRAYL) A long dirt road from Missouri to Oregon that was followed by pioneers.

oxen (AHK-suhn) A type of cattle that the pioneers used for pulling their wagons.

pioneer (pie-uh-NEAR) A person who settles in an area where very few people have lived before.

subtraction (sub-TRAK-shun) The taking away of one number from another number to find the amount left over.

supplies (suh-PLYZ) Things needed to do something.

Index